Drones

Jill McDougall

Contents

Drones

Introducing Drones

A drone is a vehicle that does not have a human pilot or driver on board. Some drones are operated by a person on the ground using remote control. Others follow a preset course without a ground pilot. These drones have inbuilt computers that help control their movements.

Not all drones are **aerial** drones, or drones that fly. There are also drones that glide beneath the world's oceans, and others that can both fly and travel under water.

Flying drones are known as Unmanned Aerial Vehicles or UAVs.

an aerial drone with a camera

Drones are used to carry out a surprising number of different tasks. For example, a drone can be used to inspect a power line, deliver a package or fly over a farmer's crop to check for pests.

Underwater drones allow their pilots to explore places such as coral reefs and shipwrecks.

Drones are often used to tackle big jobs, because they can complete them quickly and save people a great deal of time and effort.

However, not all drones are put to work. Some people fly drones as a hobby. Drone racing has become a popular sport around the world.

an underwater drone

The History of Drones

One of the earliest drones was built during World War I (1914–1918). An inventor called Charles Kettering built a drone that would carry bombs to distant targets. The drone was partly built of **papier mâché** (pronounced *pa-pee-ay mash-ay*) with cardboard wings. On its first flight, Kettering's drone crashed, and it was never used in combat. However, inventors now had the idea of developing drones to be used in war.

During the twentieth century, different kinds of **military** drones were made. Governments realised that drones were cheaper to build than regular aircraft and they could be used in conflict without any danger to aircrew. Today, military drones are commonly used in warfare.

Kettering's drone was nicknamed "the Kettering Bug" because it resembled a giant mosquito.

The Kettering Bug was much larger than many modern drones.

In the early twenty-first century, the technology of drones began to advance rapidly. With powerful cameras and **sensors** on board them, drones became useful for many different activities, from farming to making movies.

By 2015, small, light drones had become popular with the general public, especially for taking aerial photos and videos. People with smartphones or tablets could now fly a drone using **Wi-Fi** and post images from the flight on **social media.**

During this time, drones that could deliver packages to customers were designed. By 2021, there were drones delivering everything from books to medical products.

Today, aerial drones resemble flying robots that can do anything from tracking hurricanes to exploring inside pyramids, or even searching for life in outer space.

Using a drone, people can take photos of themselves in beautiful places.

Types of Drones

Drones are built in many shapes and sizes. Some are like small helicopters with horizontal blades called "rotors". Each rotor is driven by a small motor.

Drones with four rotors are known as "quadcopters". These little drones are popular as hobby drones because they are simple to operate and they can be assembled from a kit. Some quadcopters are small enough to fit inside a backpack or even a pocket.

Larger drones, which are used to carry heavy items, need many rotors to allow them to fly. One giant electric drone called a VTOL (which stands for **V**ertical **T**ake **O**ff and **L**anding) has eighteen rotors and can carry passengers.

Quadcopters can be assembled from kits.

Large electric VTOL drones can carry passengers.

Some of the largest drones resemble aeroplanes, with engines, stiff wings and vertical propellers. These drones are known as "fixed-wing" drones. They can fly higher and faster than drones with rotors, and stay in the air longer.

The biggest drones are over 20 metres long and can be as high as a two-storey building. Some of these massive aircraft need runways to take off and land, just as planes do.

a large fixed-wing drone

How Drones Work

To rise into the air, aerial drones need **lift**. This is achieved when air is moved across wings and rotors.

Most fixed-wing drones use forward speed and a vertical propeller to achieve lift.

A quadcopter rises into the air when its rotors start spinning. The rotors push air downwards, which lifts the drone off the ground. As the rotors spin faster, the drone climbs higher.

Parts of a Quadcopter

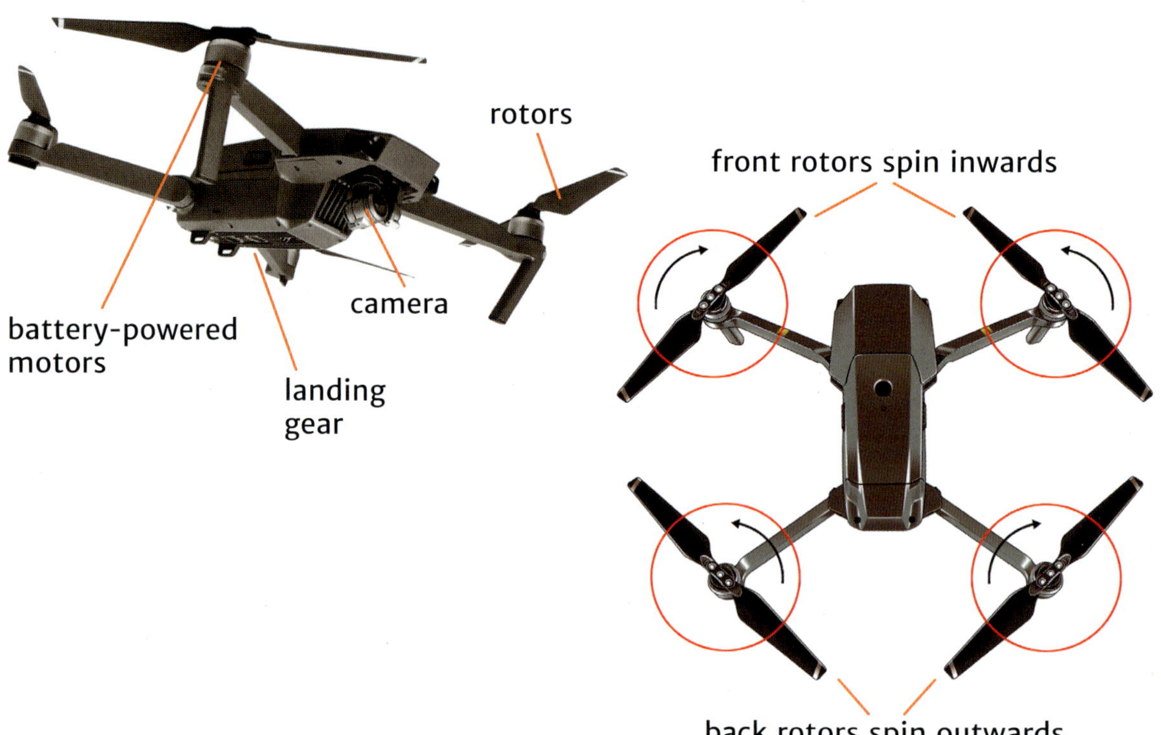

A drone can be controlled using a smartphone, a tablet or a remote-control unit. The drone pilot sends instructions to the drone using an internet connection such as Wi-Fi or **Bluetooth**.

Remote-control units are similar to gaming controllers. They usually have joysticks and buttons that are used to direct the drone's movements.

A drone pilot can often view **live video** from the drone's camera. This allows the pilot to see what the drone camera sees at the same moment.

A smartphone connected to a drone's remote-control unit displays live video from the drone high above.

Some drone pilots wear a special headset to view live video. This headset is called FPV (first-person view) goggles.

A quadcopter can perform four main flight movements. These are "throttle", "roll", "yaw" and "pitch".

Throttle moves the drone up and down. This happens when the drone pilot changes the speed of all rotors, so more or less air is pushed towards the ground.

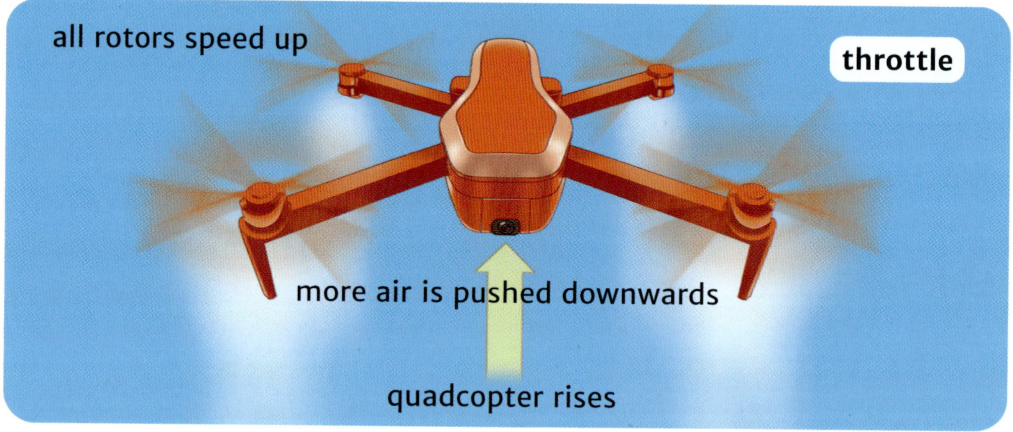

Roll moves the drone left or right. The pilot speeds up the rotors on one side and slows the rotors on the other side.

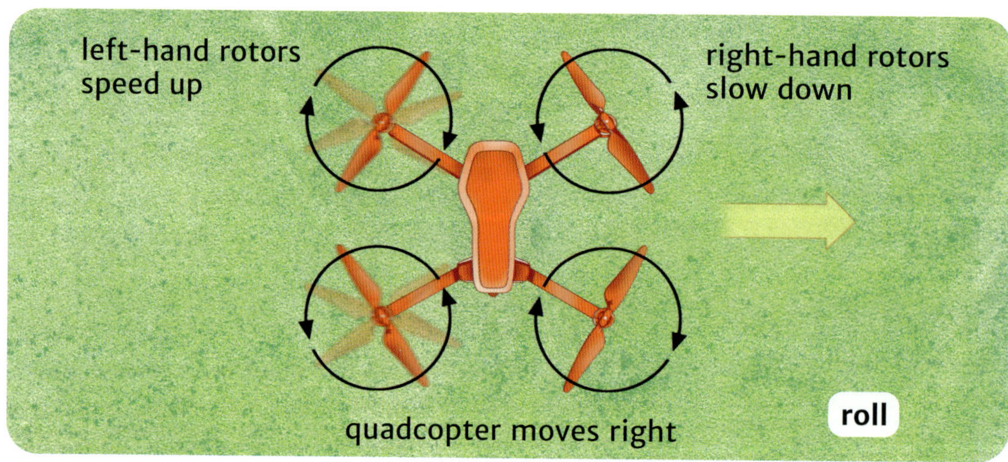

Yaw rotates the drone, or turns it around. The pilot speeds up the rotors that are spinning in one direction and slows down the rotors that are spinning in the opposite direction.

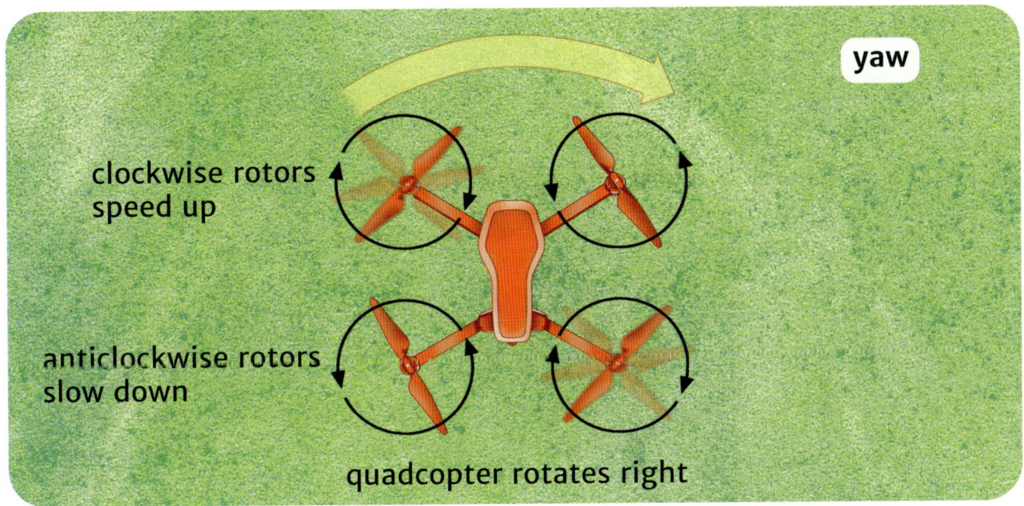

Pitch tilts the front of the drone down or up and makes it move forwards or backwards. The pilot changes the speed of the rotors at the front or back of the drone.

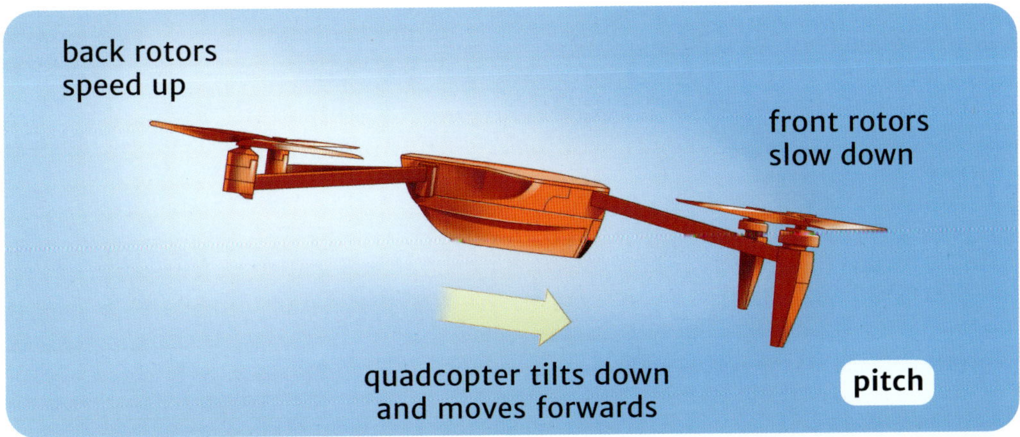

Features of Drones

Drones can perform many different functions using cameras and sensors. Drone cameras allow the pilot to capture photos or record videos. Some drone cameras are able to zoom in on tiny details when the drone is high overhead.

Scenes in movies are often filmed using drones. Filmmakers once used helicopters for filming outdoor scenes, but drones are now commonly used. Unlike helicopters, drones can fly though tiny spaces, following actors as they run through buildings or under bridges.

Different sensors on drones help them to perform many functions. **Navigation** sensors allow the drone to follow a set **flight path**. These sensors can also detect objects ahead such as trees and buildings, and change the drone's direction to avoid a crash.

To detect the warm rays given off by living things as well as objects, many drones have heat sensors. These sensors can be used to find a missing person or even to measure the temperature of a volcano that is about to erupt.

an aerial photo taken by a drone

A helicopter carries a large camera for aerial filming in 1988.

A music video crew prepares to launch a drone for filming.

Some drones have other features that enable them to do particular tasks. New drones are being made with robotic arms that have claws at the ends. These claws can allow the drone to perch on a branch or a pole, or to pick up and put down packages.

Some drones that deliver items can hover above the ground and lower the package on a line that they are fitted with. These drones are often used to deliver goods to customers. In some places, they deliver pizza!

A drone delivers a pizza.

Life-Saving Drones

Drones have become vital tools in saving human lives. They are often used during disasters such as bushfires and for search-and-rescue missions.

During bushfires, drones can send firefighters an aerial view that shows the size of the fire and the direction it is moving. Rescue teams are then able to move in quickly to save people and buildings.

Drones with heat sensors can alert firefighters to small "spot fires" started by flying sparks or embers from the main fire.

Firefighters sometimes use drones to help them monitor bushfires.

After natural disasters such as earthquakes and hurricanes, drones are often used to scan collapsed buildings for survivors. Heat sensors can pick up signals from deep under the rubble and send the exact location to rescue teams.

Sometimes, after a major disaster, it is impossible for rescuers to reach an area by road or even helicopter. Drones are often used to deliver items such as food, water and medical supplies to people trapped there.

An emergency-response team sends a drone to record water levels after a flood.

A man monitors drone video footage in a disaster control centre.

In Australia, drones are used to protect swimmers and surfers from sharks. A type of drone called a Little Ripper has a computer program that can identify the difference between people, sharks and other marine life such as dolphins. If a shark is spotted, the information is sent to a drone pilot on the shore and swimmers are quickly warned of the danger.

Little Ripper drones also carry an inflatable device that can be dropped to swimmers in trouble.

Little Ripper drones can quickly carry inflatable devices to swimmers in trouble.

Using Drones to Help the Environment

Drones are changing the ways people can protect the environment. In some places, drones are used to plant trees. The drones are packed with small, round pods that contain soil, fertiliser and seeds. As the drones fly close to the ground, they release the pods.

Some drones have special equipment that can fire the pod at the ground with enough force to bury the pod in the soil.

In coastal areas where mangrove trees are disappearing due to problems such as land clearing, drones are being used to plant mangrove seeds. In less than a year, thousands of pods have sprouted into mangrove saplings.

A single drone can plant 40 000 trees a day.

How Drones Plant Trees

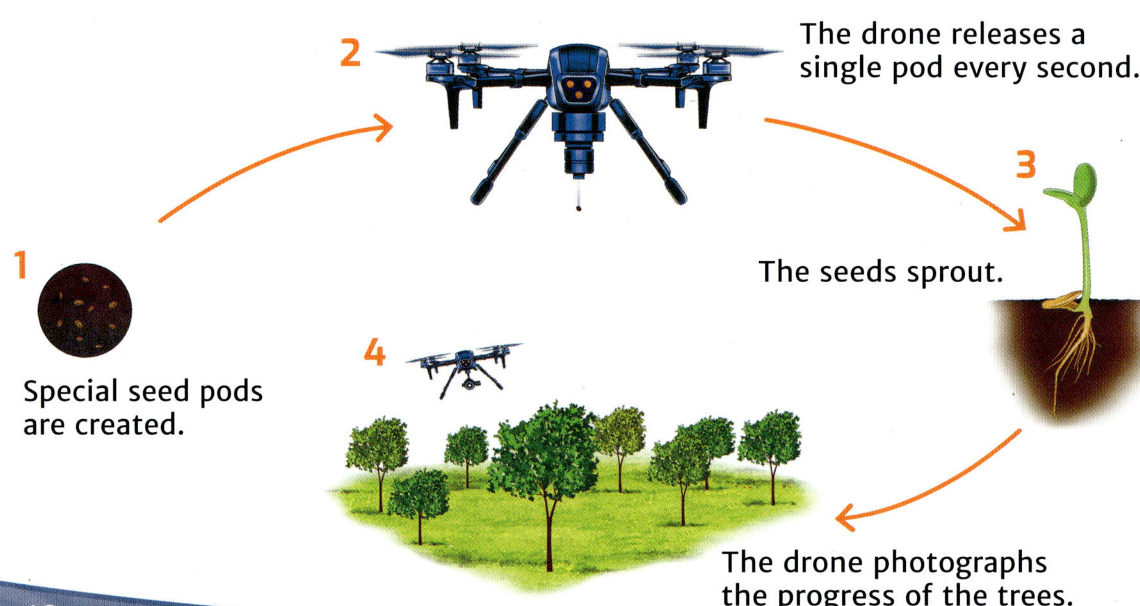

In parts of Africa, drones are being used to protect elephants. Every year, thousands of elephants are killed by **poachers** who hunt them for their tusks. Now, people are using drones to help save these endangered animals from extinction.

At night, the drones fly over the national parks and reserves where elephants roam. Drone cameras capture live video of the poachers, and map their location. This is sent to drone pilots on the ground, who then contact park rangers and police.

In some areas where drones operate, elephant poaching has stopped altogether.

Elephants can roam more safely with drones scanning for poachers.

Along the east coast of North America, scientists use drones to collect information about endangered whales.

The scientists launch a drone from a sailboat to take detailed photos of North Atlantic right whales, and also collect samples of the whales' "blow": the moist breath that a whale sprays out of its blowhole when it exhales. The photos and blow samples help scientists study the whales' health.

The North Atlantic right whale is one of the most endangered whales on the planet.

In some countries, drones are used to clean waterways of rubbish. The Cleaning Drone is an unmanned vehicle that travels around harbours and rivers, collecting rubbish before it is washed out to sea.

The drone's camera can recognise many kinds of waste, including drink containers and plastic bags.

The Cleaning Drone has a rubbish container mounted at the front. When the drone identifies rubbish floating on the water, it sails into it, forcing the rubbish to enter the bin. Once the rubbish container is full, the drone empties the waste into a larger bin and continues on its journey.

a Cleaning Drone

Underwater Drones

Underwater drones allow pilots to explore places that are too difficult or dangerous for a diver to reach. The drones' powerful cameras and lights allow drone pilots to guide the drones through the deepest and darkest waters.

Shipwreck hunters often use underwater drones to locate sunken wrecks. In 2021, shipwreck hunters used a drone to explore Lake Superior in North America. In a single month, they discovered three nineteenth-century shipwrecks. One of the wrecks lay 182 metres below the surface.

In 1985, an underwater drone discovered the wreck of the *Titanic*, which was sunk when it struck an iceberg in 1912.

A drone explores the wreck of the *Titanic*.

On the Great Barrier Reef off the coast of Queensland, Australia, underwater drones have been designed to grow coral. In some parts of the reef, the coral has died due to pollution, storms and rising sea temperatures.

Scientists load the drone with hundreds of millions of coral larvae (tiny coral that can become adult coral) and send the drone into areas where the reef has died. The drone gently squirts the coral larvae onto the seafloor and, in time, new corals grow.

The LarvalBot drone was designed to grow new coral in damaged areas of the Great Barrier Reef.

Hobby Drones

Many people enjoy flying drones just for fun. Hobby drones are often used to capture overhead images of people, places and activities such as sporting events.

Most hobby drones are quadcopters, with tough frames in case they fly into objects or crash to the ground. Many of these drones can fold up for easy transportation and storage.

In some parts of the world, drone hobby groups and flying clubs have been set up. Beginners can learn how to fly a drone and pick up new tricks and techniques.

Quadcopters make good hobby drones.

Drone racing has become a popular sport for drone pilots. Racing drones with powerful motors are flown through a course at a high speed, as pilots compete to reach the finish line.

Racing pilots wear video goggles linked to a camera on the front of the drone so they can see the view from the drone.

A drone-racing course will include many twists and turns as the drones fly over obstacles and through tight tunnels. Racers in competitions need to have very fast reaction times to avoid crashing their drones.

Some drone racers compete in major championships held at outdoor parks or football stadiums.

The rotors on racing drones form an "H" shape to increase speed.

Video goggles are needed to follow the action at a drone race.

Drones in Space

In April 2021, a tiny drone named Ingenuity became the first drone to fly across the surface of another planet. Ingenuity was part of a **NASA** space mission, and travelled to Mars attached to a spacecraft.

The air on Mars is not very **dense**, so Ingenuity's rotors had to spin very quickly to achieve lift in the thin atmosphere.

Ingenuity hovered in the air for a few moments before landing safely. It proved to scientists that flight on another planet was possible.

Ingenuity weighs only 1.8 kilograms and is solar-powered.

an artist's imagining of Ingenuity flying on Mars

Scientists at NASA are designing a larger drone that will fly on Titan, which is Saturn's biggest moon.

The drone, named Dragonfly, is expected to arrive on Titan around the year 2034. It will hover over Titan's mountains and rivers searching for signs of life. NASA scientists hope that this mission will provide clues to how life began on Earth.

Saturn and its moon Titan

an artist's imagining of Dragonfly resting between flights on Titan

The Drone Debate: Should They Be Banned?

By Emma and Jacob

Emma's Argument

I believe drones are a very useful invention. First, drones can carry out dangerous jobs such as inspecting power lines and wind farms. This stops people putting their own lives at risk.

Second, drones are useful in emergency situations such as floods and fires. In fact, they help rescuers save lives!

Drones also help to save the environment. For example, drones can be used to track down poachers that hunt endangered animals. Surely we should do everything possible to save our wildlife?

Finally, scientists use drones to make new discoveries about our planet and outer space. Clearly, drones are an excellent invention and should not be banned.

Jacob's Argument

I think drones should be banned. To begin with, drones with cameras affect people's privacy. Nobody wants drones flying over their garden and spying through their windows!

Also, noisy drones can upset wildlife and cause them stress. Birds with babies in the nest have been known to attack drones that came too close to them.

In addition to this, drones get in the way of rescue aircraft. There are many reports of private drones flying near bushfires and forcing fire-fighting helicopters to land.

Furthermore, drones can be extremely dangerous. In one case, a jet carrying 200 passengers was almost hit by a drone, and the pilot had to abort the landing.

Clearly, drones are a danger to people as well as animals and should be banned.

From the outer reaches of space to the deepest oceans on Earth, drones are used almost everywhere. As the technology continues to improve, drones will enter even more places and carry out many more activities. Some people believe that drones bring only benefits, but others have concerns.

Glossary

aerial (*adjective*)	up in the air
Bluetooth (*noun*)	wireless technology that allows different devices to communicate with each other
dense (*adjective*)	thick and tightly packed
flight path (*noun*)	the route an aircraft follows to reach its destination
lift (*noun*)	upward force that can raise an aircraft
live video (*noun*)	video that is viewed at the same time as the action is happening
military (*adjective*)	relating to the armed forces of a country, such as the army or navy
NASA (*noun*)	the National Aeronautics and Space Administration in the USA
navigation (*noun*)	planning a route for a vehicle and guiding it there
papier mâché (*noun*)	a mixture of paper and glue that becomes hard when dry
poachers (*noun*)	people who illegally kill or catch animals

sensors (*noun*) devices that detect and measure where and how much of something there is, such as heat or motion

social media (*noun*) websites and apps that allow users to create and share content with other users

Wi-Fi (*noun*) a wireless communication network that allows devices to connect to the internet

Index